Collins

easy le

Maths

quick quizzes

Ages 5–7

2 ? 8 9 7 5 4

Brad Thompson

Number bonds to 20

Complete the number bond.

Use the number line to help you.

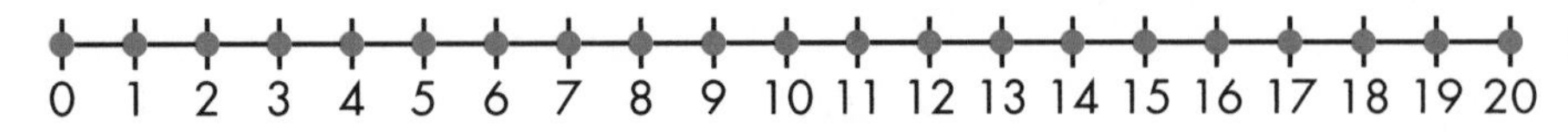

Learn these number bonds. They will help you with mental maths.

1. 0 + 10 = ☐
2. 5 + 15 = ☐
3. 10 + 10 = ☐
4. 5 + 5 = ☐
5. 20 + 0 = ☐
6. 9 + ☐ = 10
7. ☐ + 8 = 10
8. 7 + ☐ = 20
9. ☐ + 6 = 20
10. 4 + ☐ = 20
11. 20 = 18 + ☐
12. 10 = ☐ + 1
13. 20 = 8 + ☐
14. 18 = ☐ + 6
15. 20 = 1 + ☐

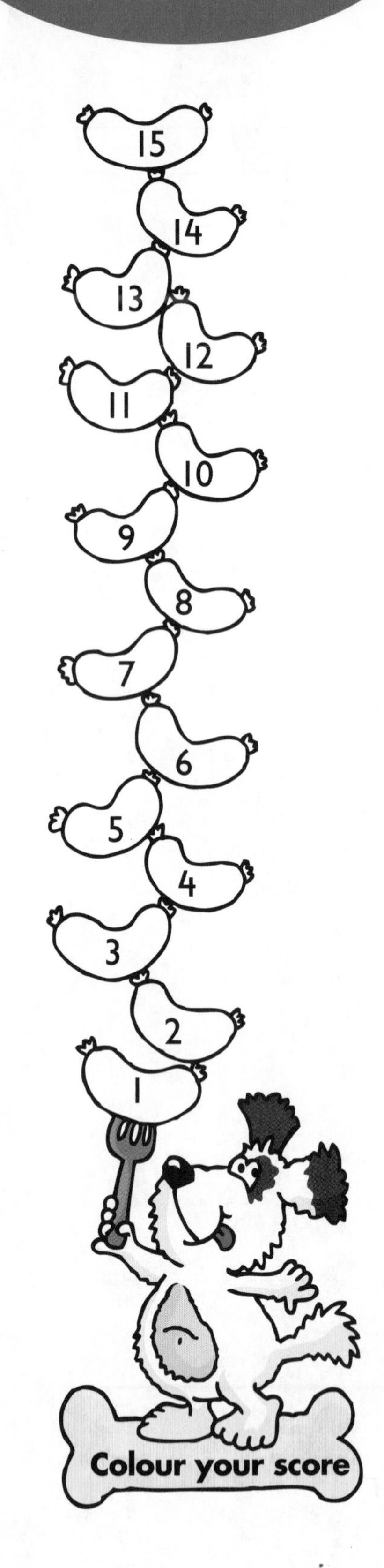

Addition

Fill in the missing numbers.

To add, start with the larger number and count on from it.

1. 7 + 2 = ☐
2. 6 + 5 = ☐
3. 10 + 3 = ☐
4. 9 + 5 = ☐
5. 5 + 2 = ☐
6. 3 + 6 = ☐
7. 9 + 10 = ☐
8. 7 + 5 = ☐
9. 4 + 4 = ☐
10. 18 + 1 = ☐

11. ☐ = 5 + 1
12. 14 = ☐ + 7
13. 16 = 3 + ☐
14. 9 = ☐ + 8

15. ☐ = 4 + 14

Find the difference

Write the difference between each pair of numbers.

Start with the smaller number and count on.

1	8	⟷	1	
2	2	⟷	10	
3	5	⟷	15	
4	10	⟷	5	
5	12	⟷	11	
6	17	⟷	13	
7	14	⟷	18	
8	16	⟷	11	
9	7	⟷	19	
10	23	⟷	12	
11	21	⟷	17	
12	23	⟷	11	
13	12	⟷	26	
14	20	⟷	35	
15	34	⟷	15	

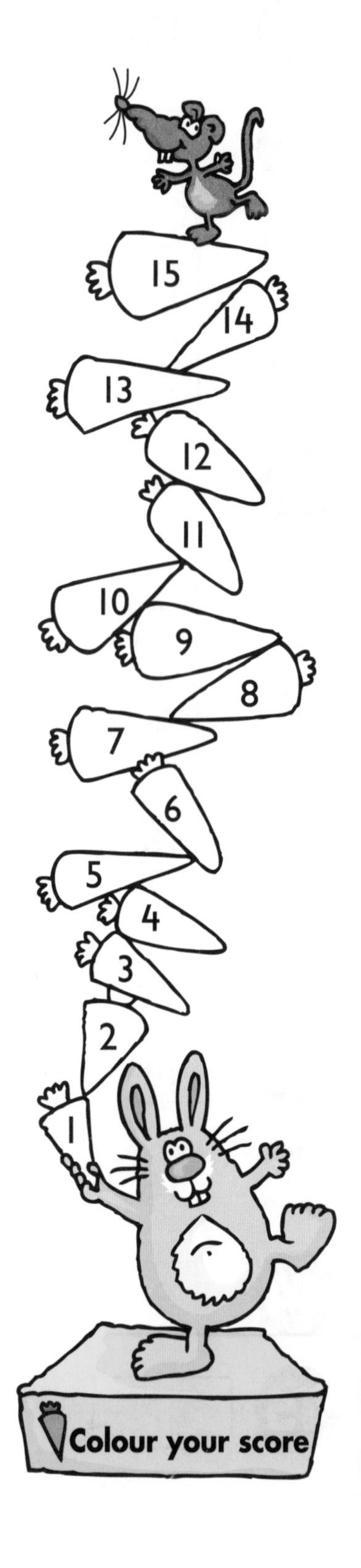

Subtraction

Complete the subtractions.

1. 25 – 2 = ☐
2. 37 – 5 = ☐
3. 49 – 3 = ☐
4. 59 – 8 = ☐
5. 64 – 2 = ☐
6. 40 – 4 = ☐
7. 50 – 6 = ☐
8. 60 – 8 = ☐
9. 70 – 7 = ☐
10. 80 – 5 = ☐
11. 23 – 9 = ☐
12. 35 – 6 = ☐
13. 41 – 9 = ☐
14. 55 – 7 = ☐
15. 61 – 3 = ☐

Count backwards from the larger number when subtracting.

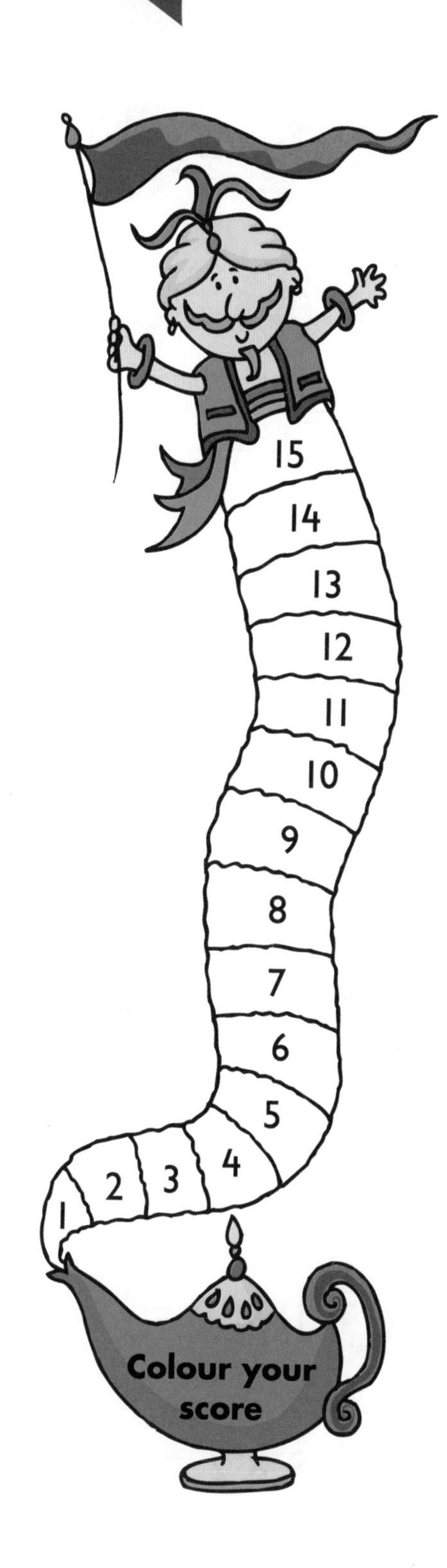

2 times table

Count the coins worth two pence and complete the number sentence.

For the two times table, count up in twos.

1.  = ☐ p

2.  = ☐ p

3. + + + = ☐ p

4. + + + +
 = ☐ p

5. + + + + +
 = ☐ p

Fill in the missing numbers.

6	3 × 2 = ☐	11	☐ × 2 = 4
7	6 × 2 = ☐	12	☐ × 2 = 8
8	1 × 2 = ☐	13	☐ × 2 = 16
9	11 × 2 = ☐	14	☐ × 2 = 20
10	8 × 2 = ☐	15	☐ × 2 = 24

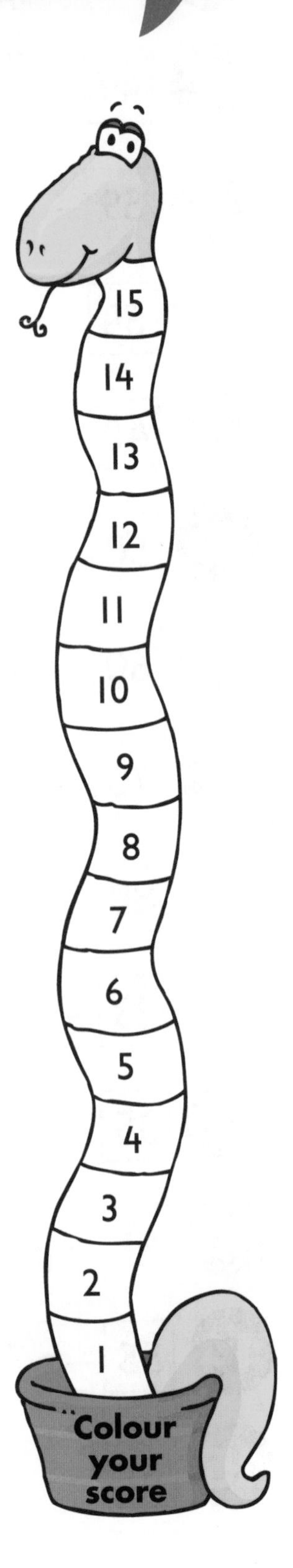

5 times table

For the five times table, count up in fives.

Count the coins worth five pence and complete the number sentence.

1. 5p + 5p = ☐ p

   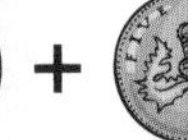

2. 5p + 5p + 5p = ☐ p

3. 5p + 5p + 5p + 5p = ☐ p

  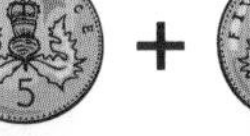   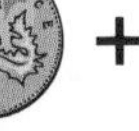 

4. 5p + 5p + 5p + 5p + 5p = ☐ p

   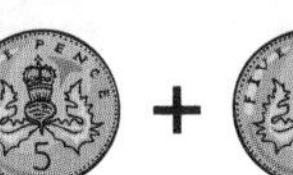

5. 5p + 5p + 5p + 5p + 5p + 5p = ☐ p

Fill in the missing numbers.

6. 3 × 5 = ☐
7. 12 × 5 = ☐
8. 1 × 5 = ☐

9. 11 × 5 = ☐
10. 8 × 5 = ☐
11. ☐ × 5 = 20

12. ☐ × 5 = 5

13. ☐ × 5 = 30

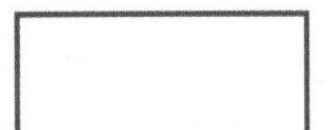

14. ☐ × 5 = 55
15. ☐ × 5 = 40

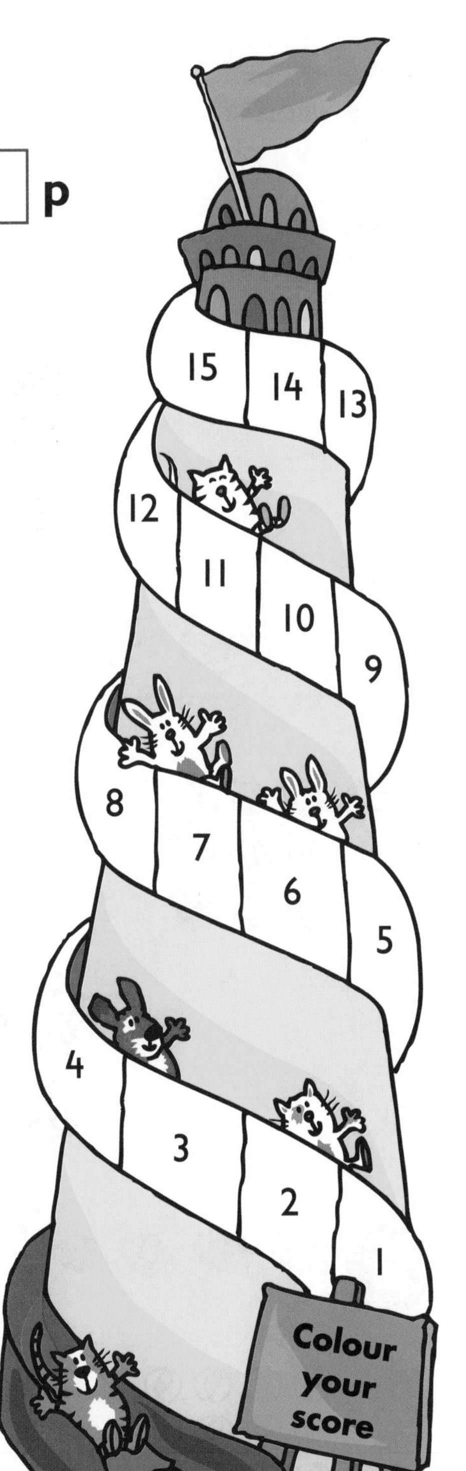

Sharing

Share the objects equally between the children. Write how many each child gets.

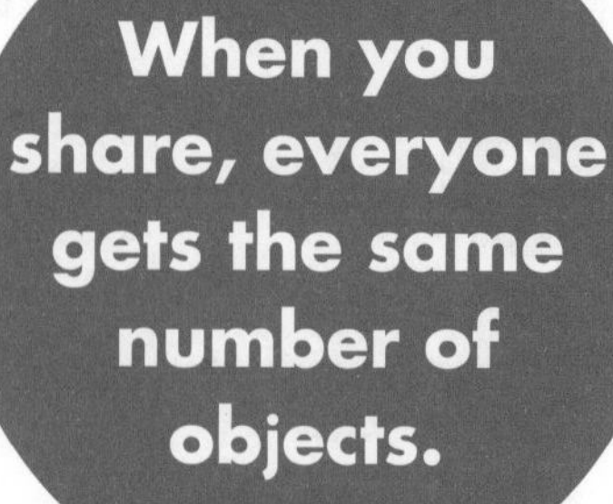

1  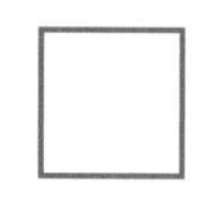

3  

2 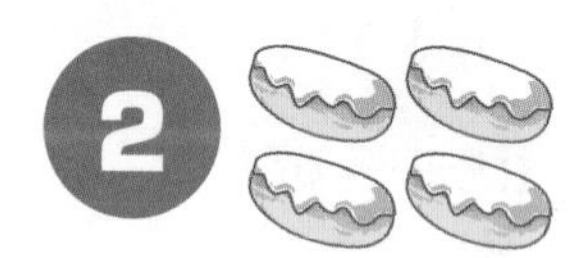 

4  

5 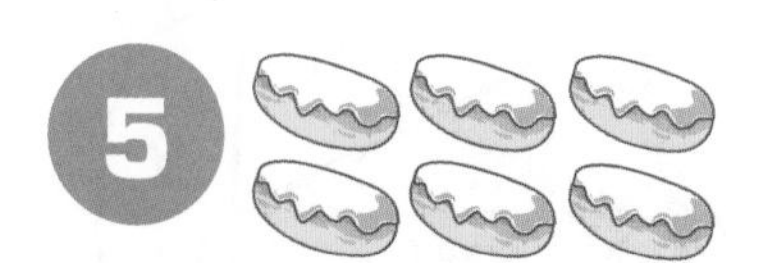 

7 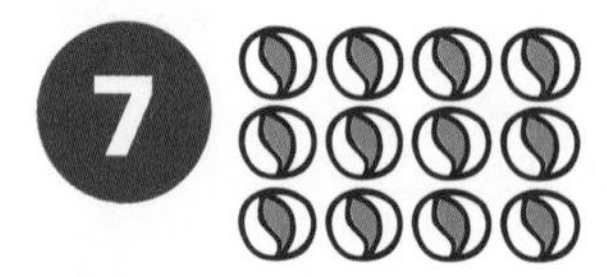 

6  

8  

9  

11 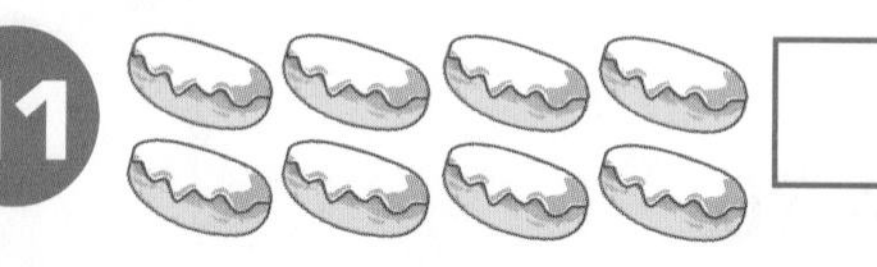

10  

12 

Dividing by 2

How many groups of two are there?

Group the objects into twos and count the groups.

1  ☐

2  ☐

3  ☐

4 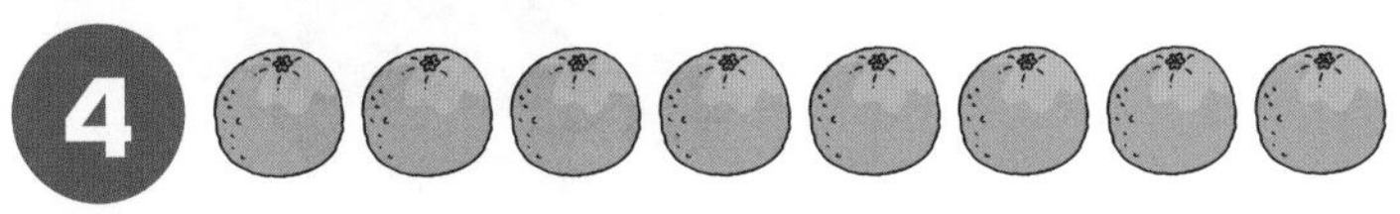 ☐

5 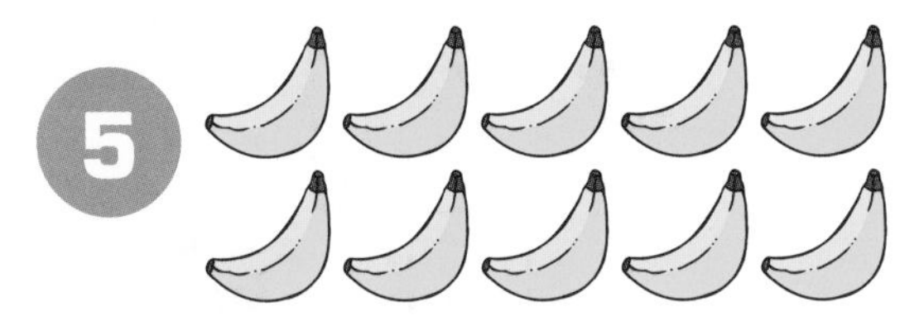 ☐

Write the answers.

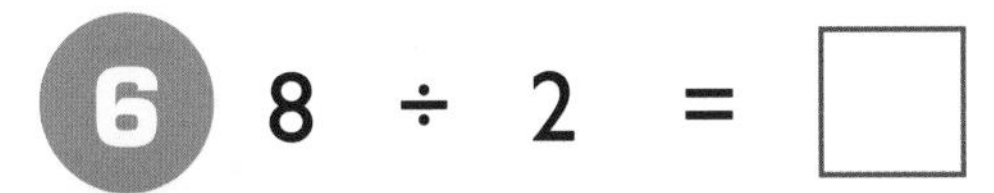

6 $8 \div 2 =$ ☐

7 $4 \div 2 =$ ☐

8 $10 \div 2 =$ ☐

9 $6 \div 2 =$ ☐

10 $2 \div 2 =$ ☐

11 18 shared by 2 → ☐

12 12 shared by 2 → ☐

13 20 shared by 2 → ☐

14 16 shared by 2 → ☐

15 14 shared by 2 → ☐

Estimating

Estimate the number shown by each arrow.

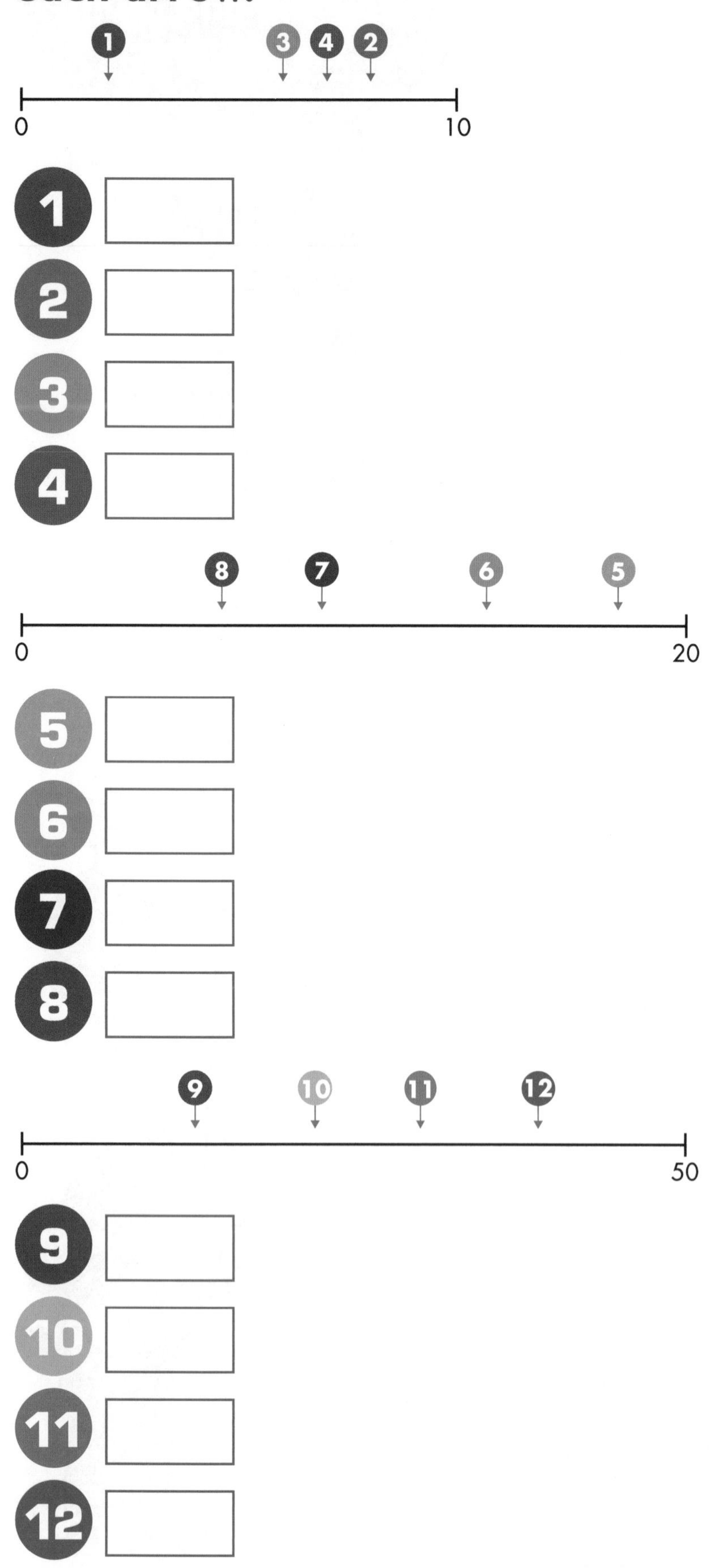

Work out the middle number and mark it on the line before you begin.

Comparing numbers

Write <, > or = in each box.

< means 'is less than'
> means 'is greater than'
= means 'is equal to'

1. 6 ☐ 7
2. 10 ☐ 9
3. 15 ☐ 20
4. 22 ☐ 19
5. 30 ☐ 25

Write any correct number to finish each number sentence.

6. 35 < ☐
7. 41 > ☐
8. 47 < ☐
9. 55 > ☐
10. 64 < ☐

Write <, > or = in each box.

11. 70 + 5 ☐ 82
12. 73 + 7 ☐ 80
13. 82 + 3 ☐ 90
14. 95 ☐ 91 + 5
15. 100 ☐ 90 + 10

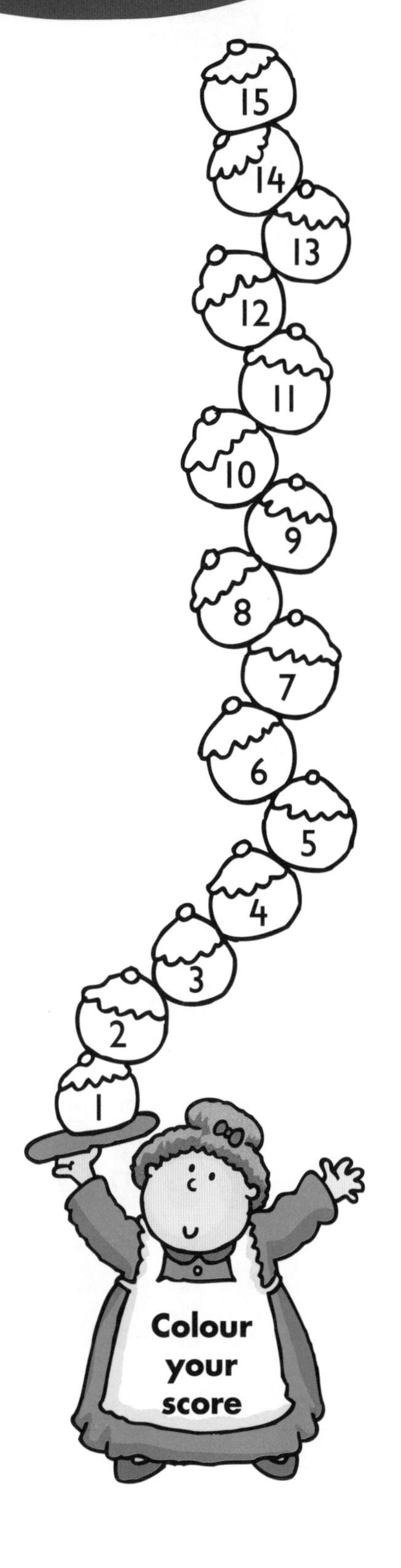

Halves

A half is one of two equal parts.

Use a tick or cross to show if the object has been halved.

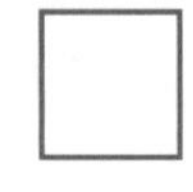

2

Colour one half.

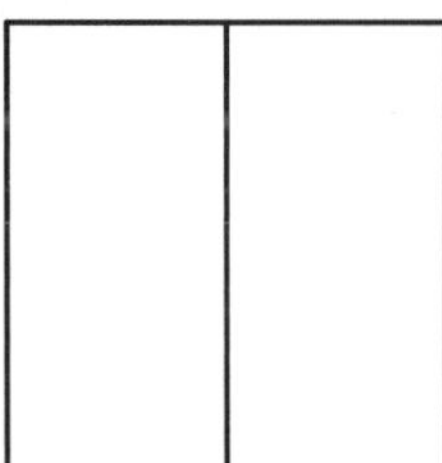

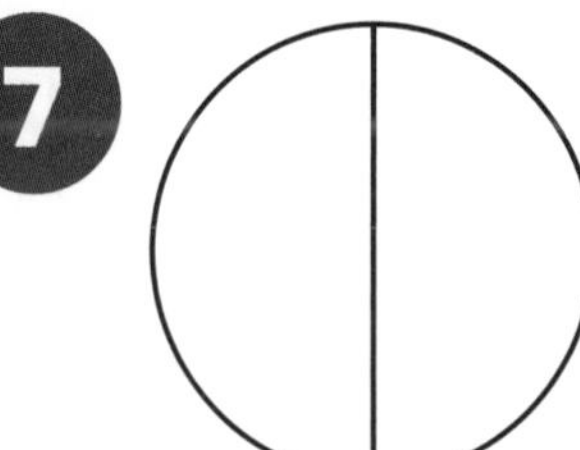

6

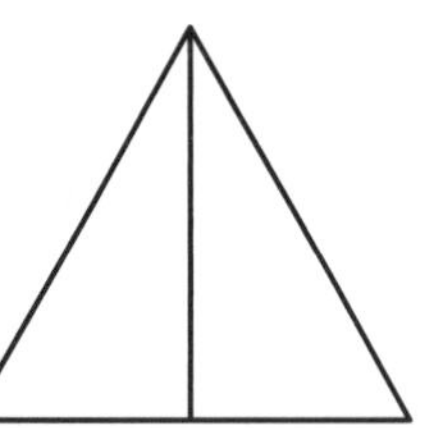

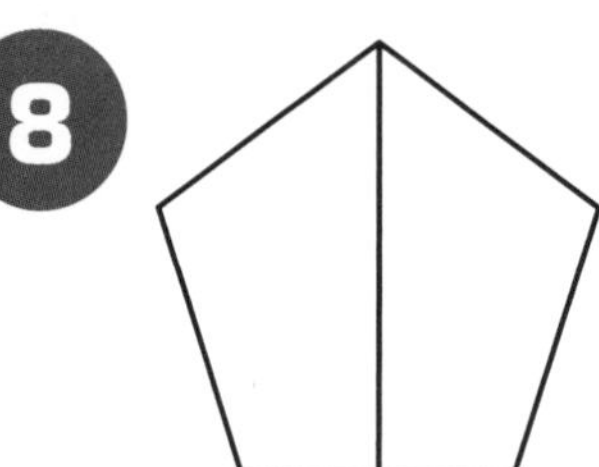

Find half of the amounts.

pencils

10

book

11

bottles

12

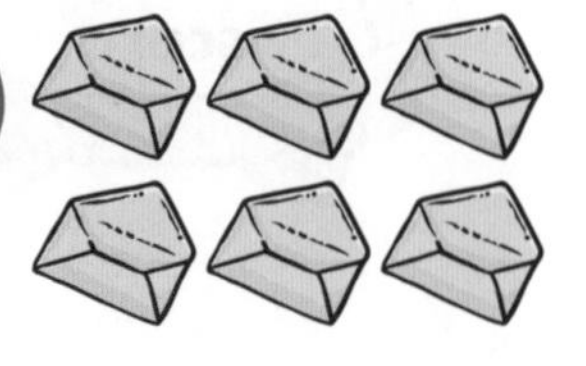

envelopes

12 11 10 9 8 7 6 5 4 3 2 1

Colour your score

Quarters

Use a tick or cross to show if the object has been divided into quarters.

A quarter is one of four equal parts.

 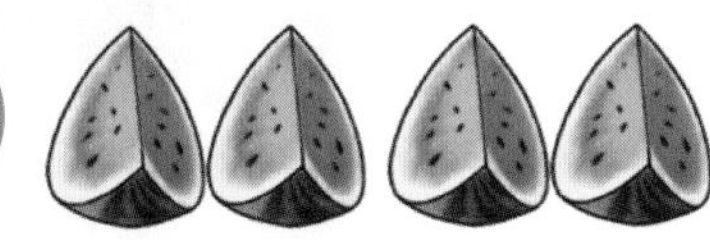 

 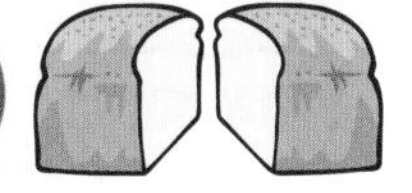 

Colour one quarter.

5 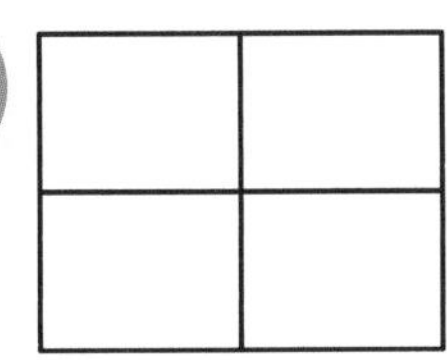

6 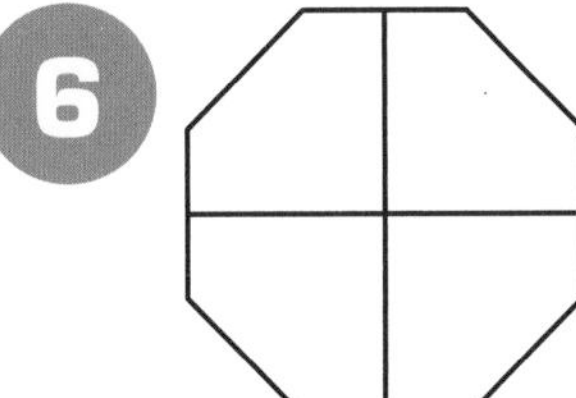

Find a quarter of the amounts.

7  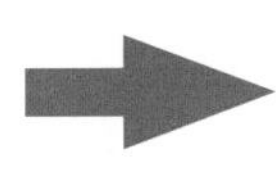  stamps

8 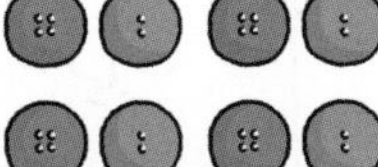 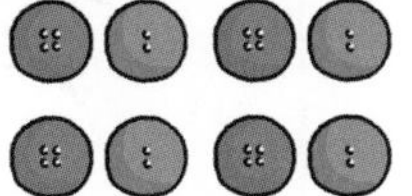 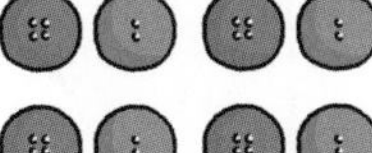  ☐ buttons

9  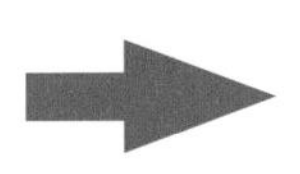 ☐ marbles

10   ☐ candles

Height and length

Use a ruler to measure the length of each crayon in centimetres.

Place 0 on your ruler at the beginning of what you are measuring.

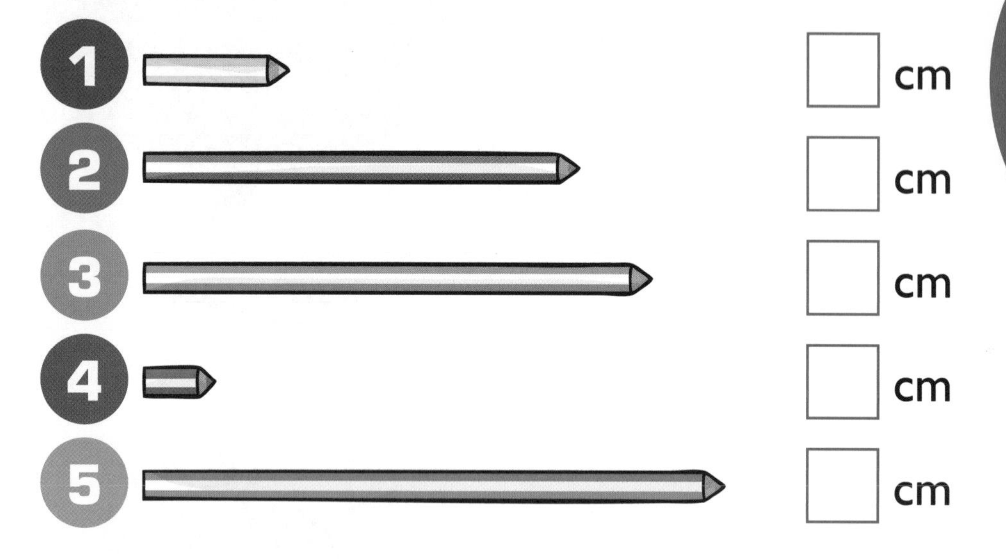

1 ☐ cm

2 ☐ cm

3 ☐ cm

4 ☐ cm

5 ☐ cm

Measure the height of each stack in cm using a ruler.

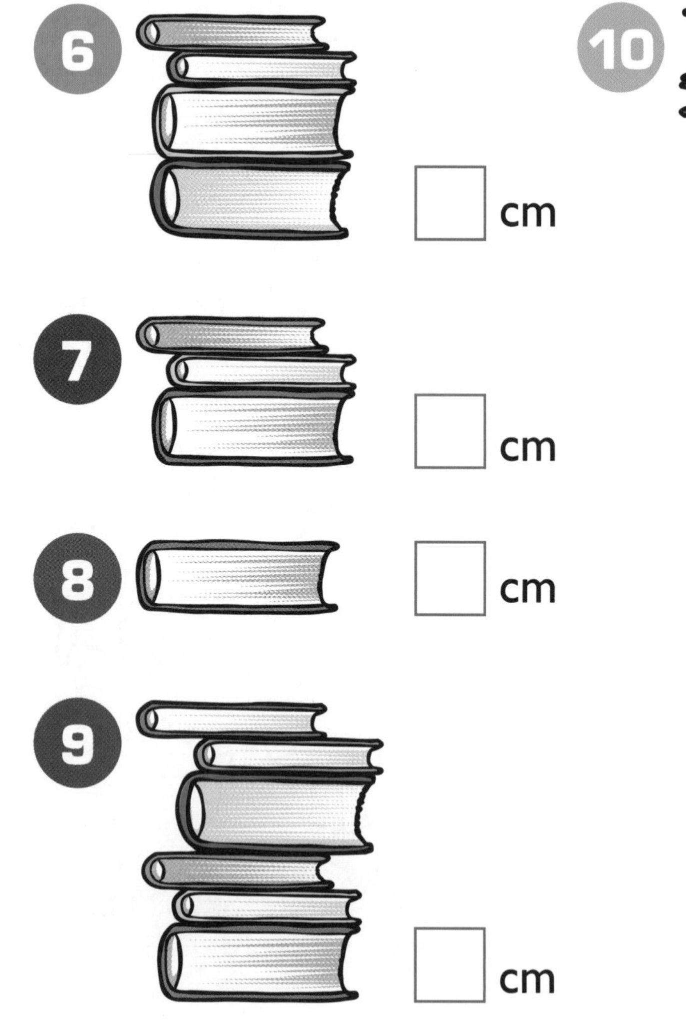

6 ☐ cm

7 ☐ cm

8 ☐ cm

9 ☐ cm

10 ☐ cm

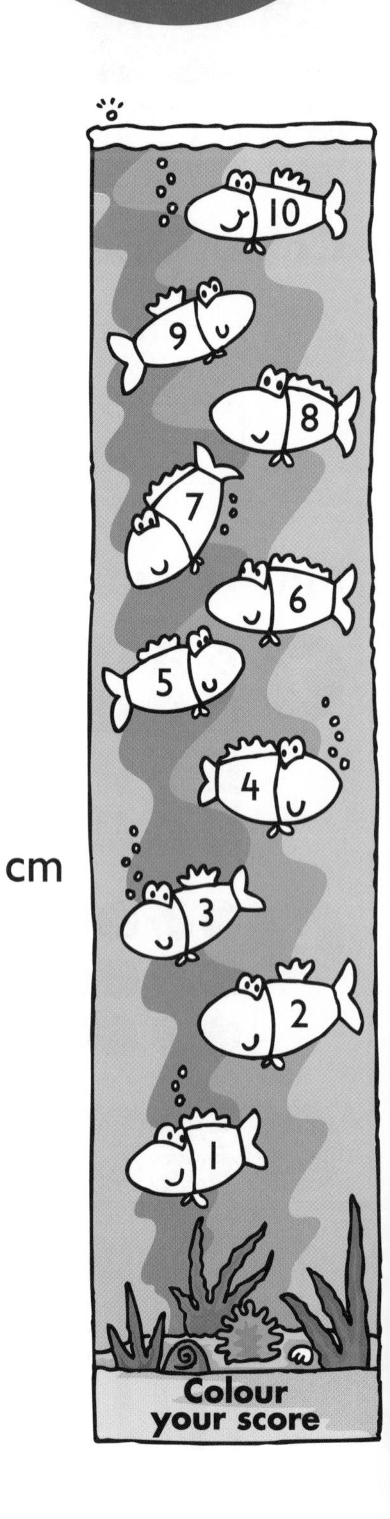

Telling the time

The short hand shows the hour. The long hand shows the minutes.

Write the time in words.

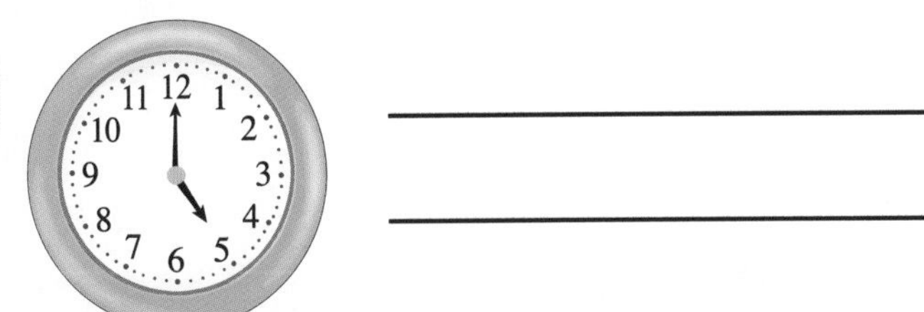

2
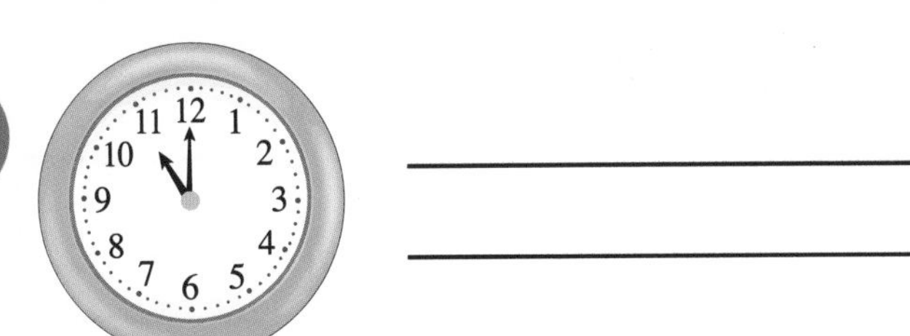

3
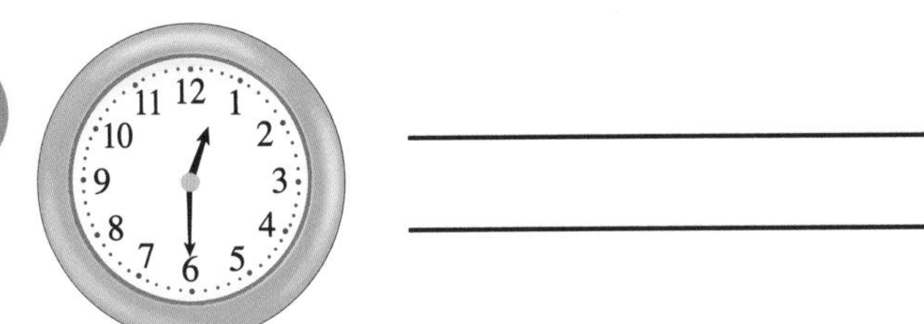

Draw the hands on the clock.

 2 o'clock

7 o'clock

6 12 o'clock

8 6 o'clock

Naming shapes

Draw a line from each shape to its name.

Two-dimensional shapes are flat. Three-dimensional shapes are solid.

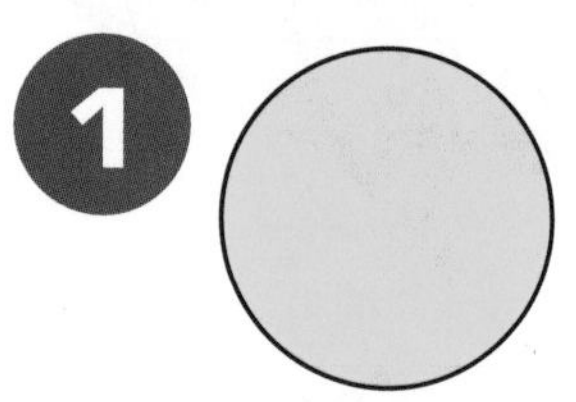

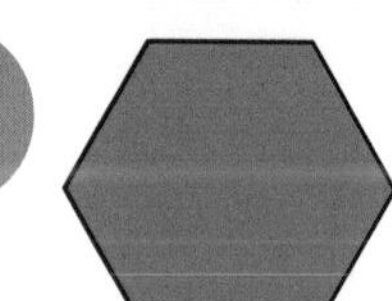

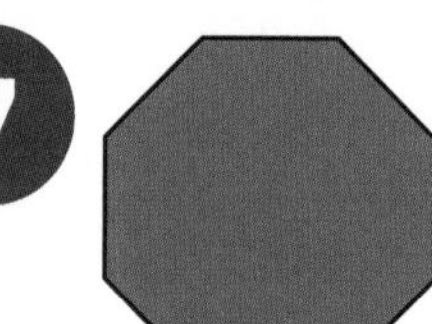

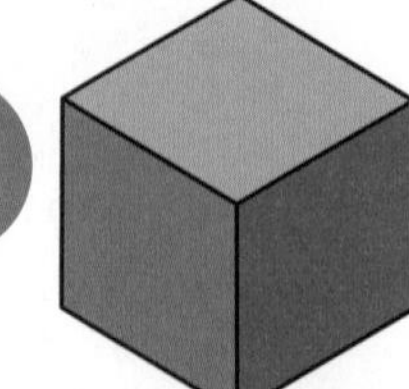
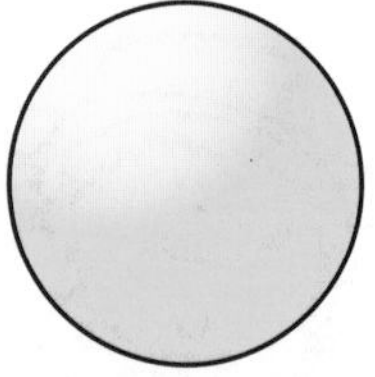
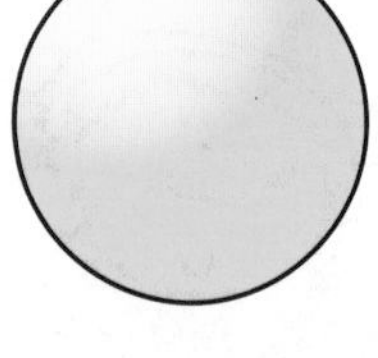

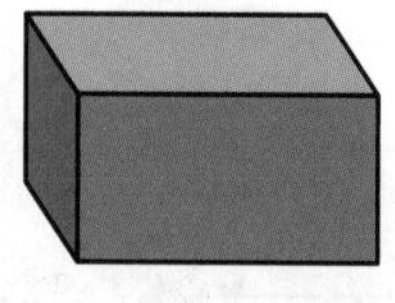

1 2 3 4 5 6 7 8

kite

circle

rectangle

pentagon

triangle

octagon

square

hexagon

9 10 11 12

cylinder

cuboid

sphere

cube

12
11
10
9
8
7
6
5
4
3
2
1

Number bonds to 100

Complete these number bonds.

Use the number bonds to 10 to help you.

1. $3 + 7 = 10$ → $30 + \square = 100$
2. $1 + 9 = 10$ → $10 + \square = 100$
3. $5 + 5 = 10$ → $50 + \square = 100$
4. $2 + 8 = 10$ → $20 + \square = 100$
5. $4 + 6 = 10$ → $40 + \square = 100$
6. $0 + 10 = 10$ → $0 + \square = 100$
7. $9 + 1 = 10$ → $90 + \square = 100$
8. $10 + 0 = 10$ → $100 + \square = 100$
9. $6 + 4 = 10$ → $60 + \square = 100$
10. $8 + 2 = 10$ → $80 + \square = 100$
11. $10 = 5 + 5$ → $100 = 50 + \square$
12. $10 = 0 + 10$ → $100 = 0 + \square$
13. $10 = 9 + 1$ → $100 = 90 + \square$
14. $10 = 2 + 8$ → $100 = 20 + \square$
15. $10 = 10 + 0$ → $100 = 100 + \square$

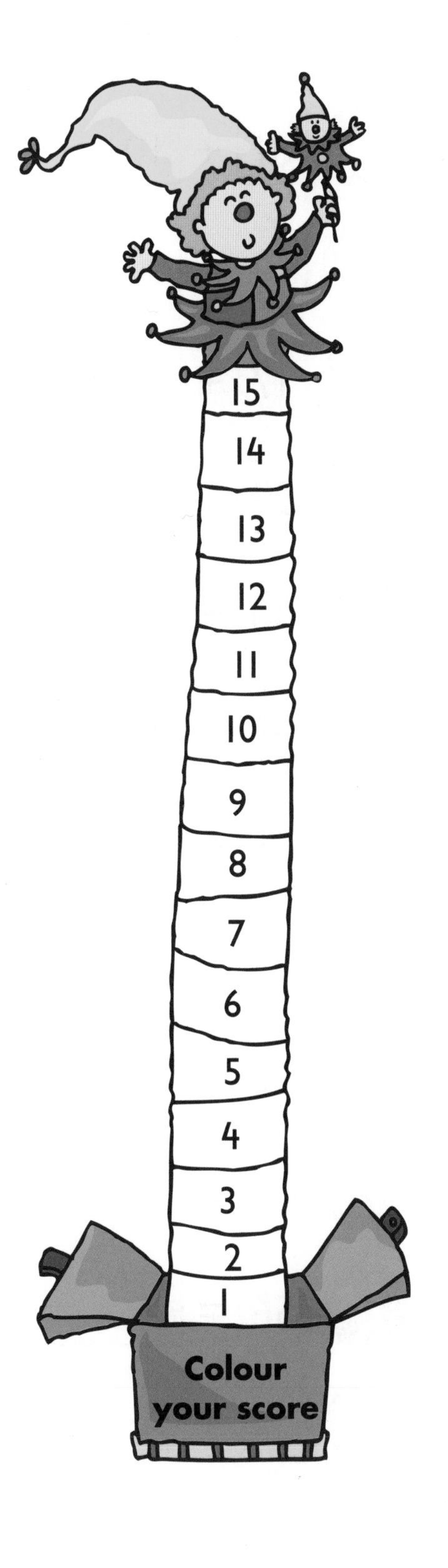

More addition

Fill in the missing numbers.

1. 37 + 10 = ☐
2. 49 + 10 = ☐
3. 55 + 10 = ☐
4. 61 + 10 = ☐
5. 74 + 10 = ☐
6. 17 + 12 = ☐
7. 23 + 15 = ☐
8. 30 + 17 = ☐
9. 48 + 11 = ☐
10. 51 + 19 = ☐
11. ☐ = 25 + 41
12. 57 = ☐ + 35
13. 43 = 15 + ☐
14. 56 = ☐ + 32
15. ☐ = 25 + 73

Add the tens, then add the ones.

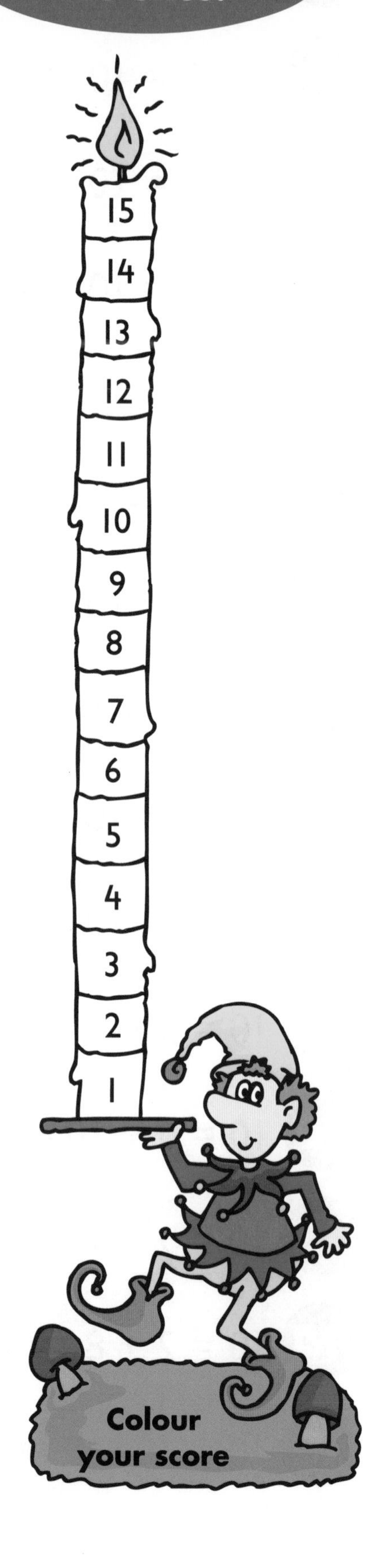

Using number bonds

Complete these number bonds.

Check where the = sign is!

1. 100 – 90 = ☐
2. 100 – 50 = ☐
3. 100 – 20 = ☐
4. 100 – 10 = ☐
5. 100 – 60 = ☐
6. 100 – 80 = ☐
7. 100 – 30 = ☐
8. 100 – 100 = ☐
9. 100 – 40 = ☐
10. 100 – 70 = ☐
11. 70 = 100 – ☐
12. 80 = 100 – ☐
13. 20 = 100 – ☐
14. 10 = 100 – ☐
15. 50 = 100 – ☐

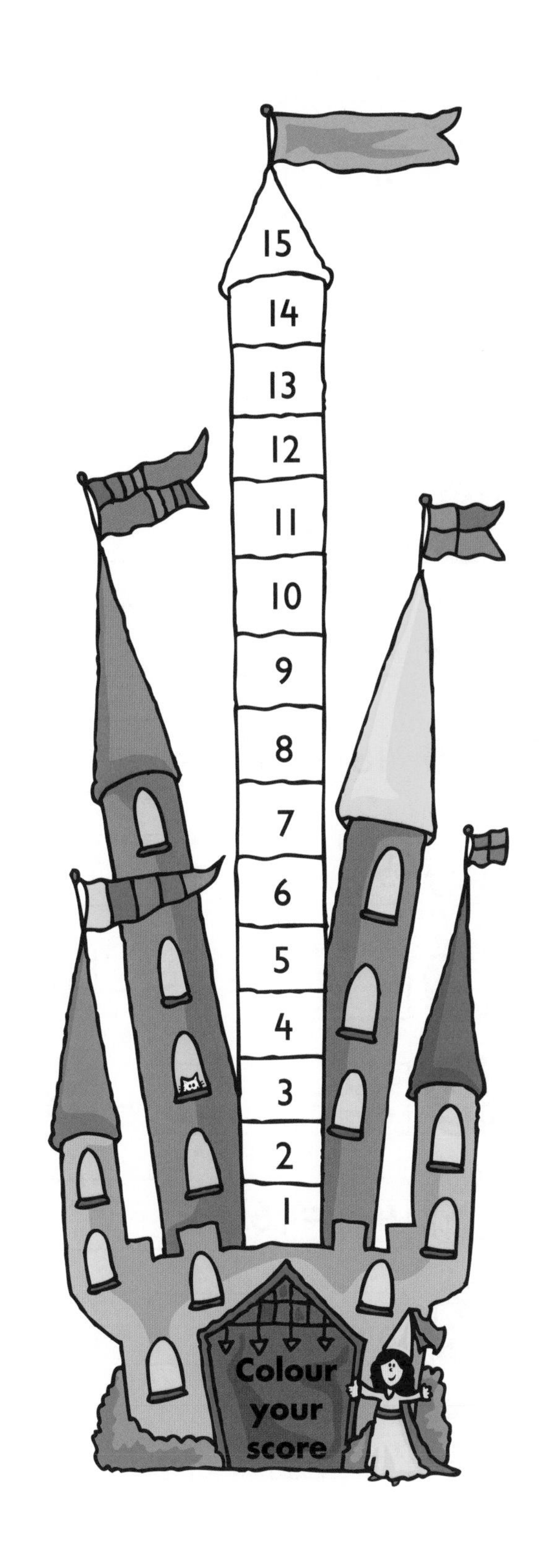

More subtraction

Fill in the missing numbers.

1. 17 – 10 = ☐
2. 29 – 10 = ☐
3. 35 – 10 = ☐
4. 41 – 10 = ☐
5. 51 – 10 = ☐
6. 17 – 12 = ☐
7. 23 – 15 = ☐
8. 30 – 17 = ☐
9. 48 – 11 = ☐
10. 52 – 19 = ☐
11. ☐ = 25 – 11
12. 57 = ☐ – 13
13. 43 = 65 – ☐
14. 56 = ☐ – 22
15. ☐ = 65 – 43

Subtracting from a number will always give a smaller number.

10 times table

Count the coins worth ten pence and complete the number sentence.

For the ten times table, count up in tens.

1 
10 + 10 = ☐ p

2 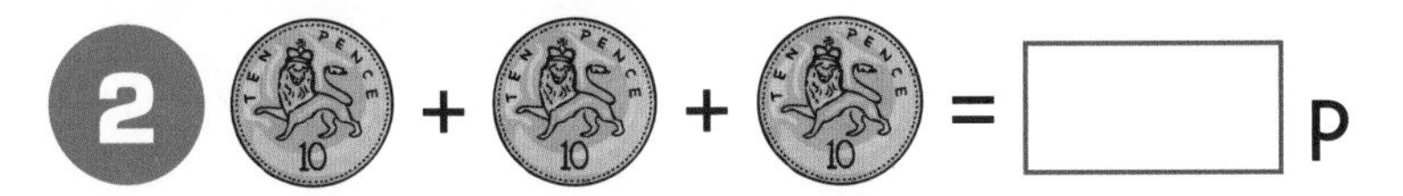
10 + 10 + 10 = ☐ p

3 
10 + 10 + 10 + 10 = ☐ p

4 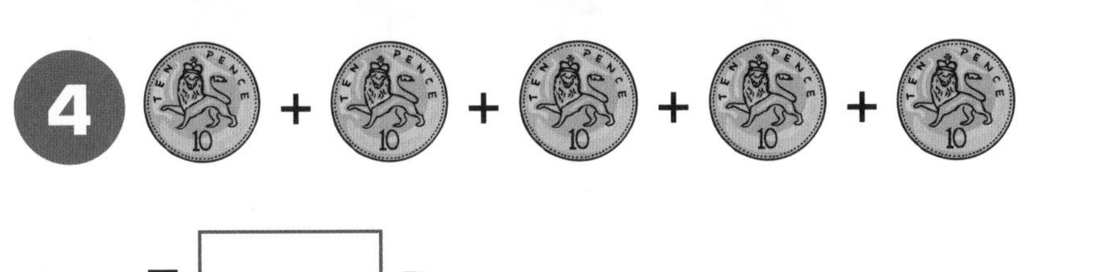
10 + 10 + 10 + 10 + 10 = ☐ p

5 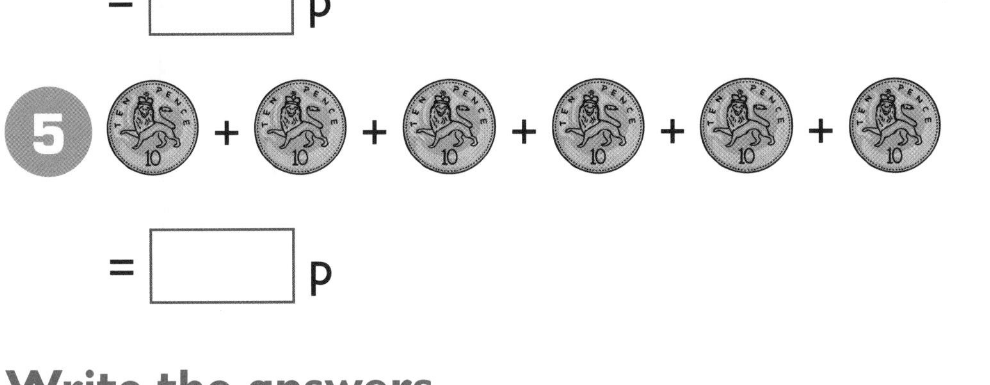

10 + 10 + 10 + 10 + 10 + 10 = ☐ p

Write the answers.

6 $3 \times 10 =$ ☐

7 $6 \times 10 =$ ☐

8 $1 \times 10 =$ ☐

9 $11 \times 10 =$ ☐

10 $8 \times 10 =$ ☐

11 ☐ $\times 10 = 20$

12 ☐ $\times 10 = 50$

13 ☐ $\times 10 = 30$

14 ☐ $\times 10 = 120$

15 ☐ $\times 10 = 40$

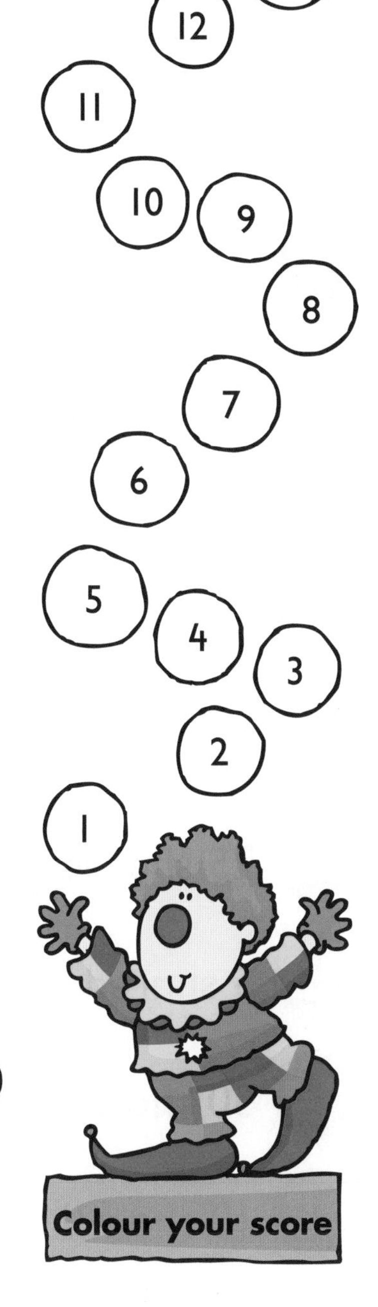

Dividing by 5

Fill in the missing numbers.

1. 5 ÷ 5 = ☐
2. 10 ÷ 5 = ☐
3. 15 ÷ 5 = ☐
4. 20 ÷ 5 = ☐
5. 25 ÷ 5 = ☐
6. ☐ ÷ 5 = 7
7. ☐ ÷ 5 = 9
8. ☐ ÷ 5 = 6
9. ☐ ÷ 5 = 10
10. ☐ ÷ 5 = 8
11. 60 shared by 5 → ☐
12. 25 shared by 5 → ☐
13. 5 shared by 5 → ☐
14. 50 shared by 5 → ☐
15. 10 shared by 5 → ☐

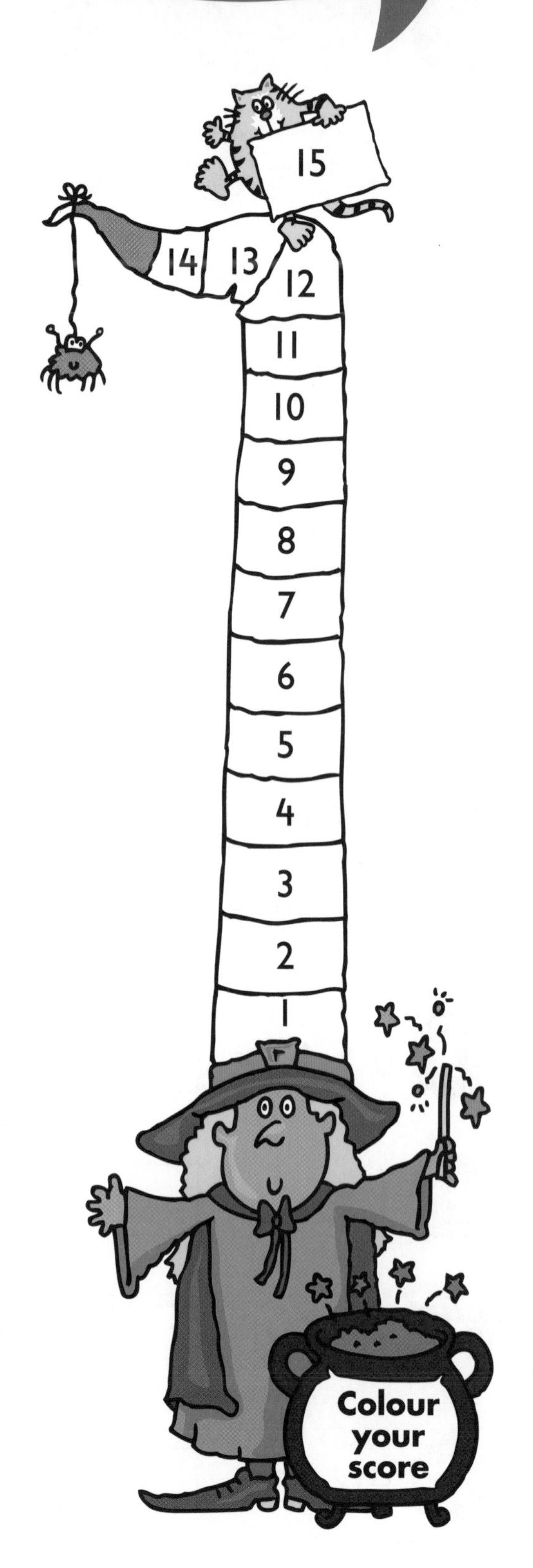

Dividing by 10

Fill in the missing numbers.

When dividing by ten, remember to group things into tens.

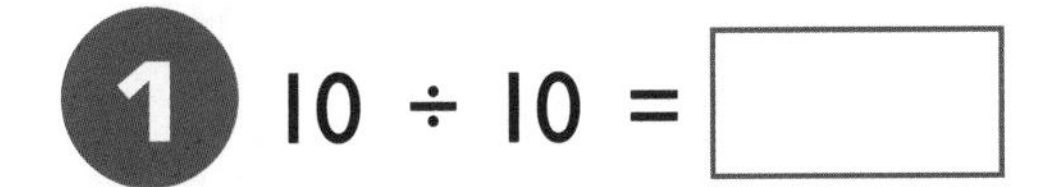

1. $10 \div 10 = \square$
2. $20 \div 10 = \square$
3. $30 \div 10 = \square$
4. $40 \div 10 = \square$
5. $50 \div 10 = \square$
6. $\square \div 10 = 4$
7. $\square \div 10 = 8$
8. $\square \div 10 = 9$
9. $\square \div 10 = 6$
10. $\square \div 10 = 11$
11. 120 shared by 10 → $\square$
12. 40 shared by 10 → $\square$
13. 50 shared by 10 → $\square$
14. 110 shared by 10 → $\square$
15. 10 shared by 10 → $\square$

Finding one third

Use a tick or a cross to show if the object has been cut into thirds.

A third is one of three equal parts.

 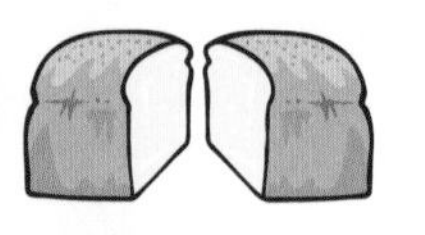 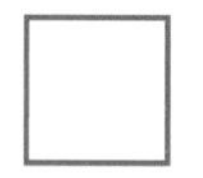

  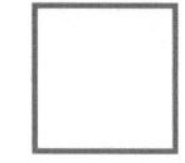

 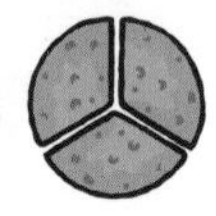 

Colour in one third.

 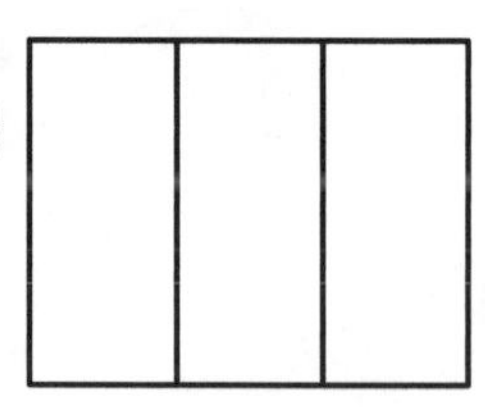

7 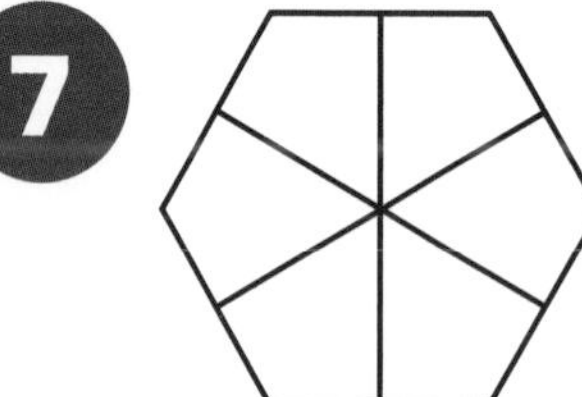

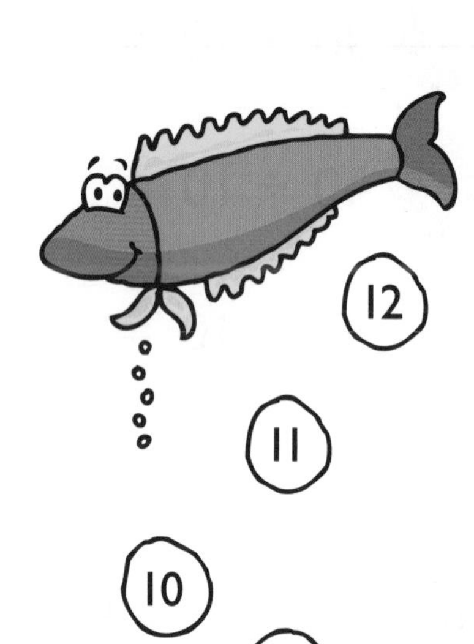

6 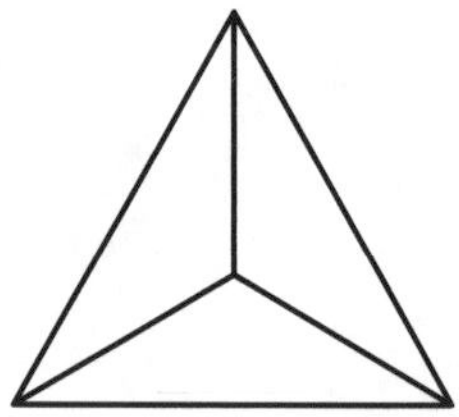

8 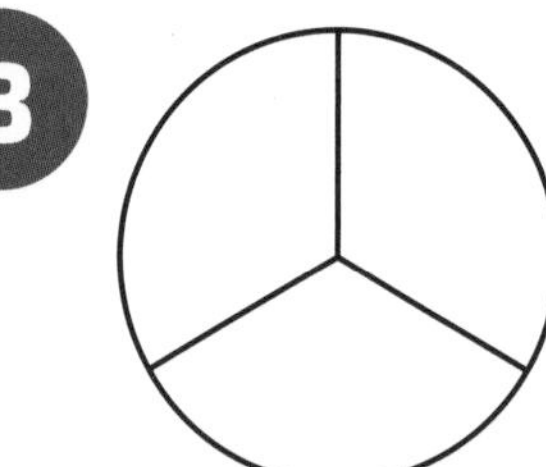

Find a third of the amounts.

9   

counters

10 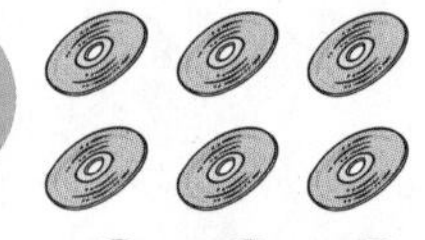  

CDs

11 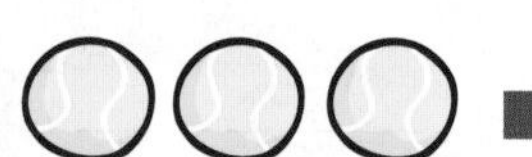  

tennis ball

 crayons

Finding three quarters

Put a tick if three quarters are shown. Put a cross if not.

 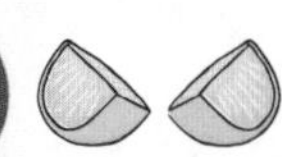 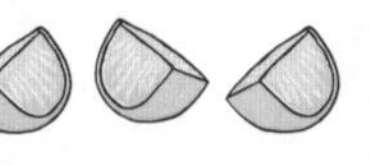 

 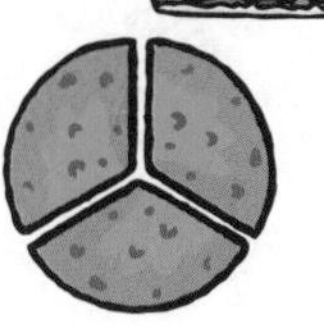

Three quarters are three of four equal parts.

Colour in three quarters.

 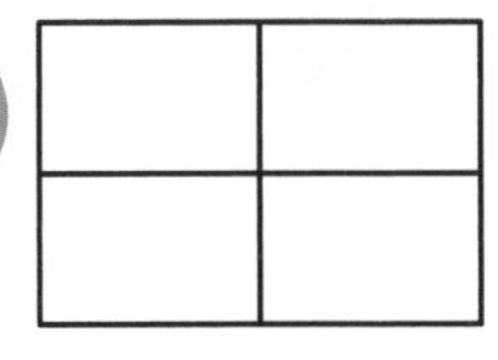

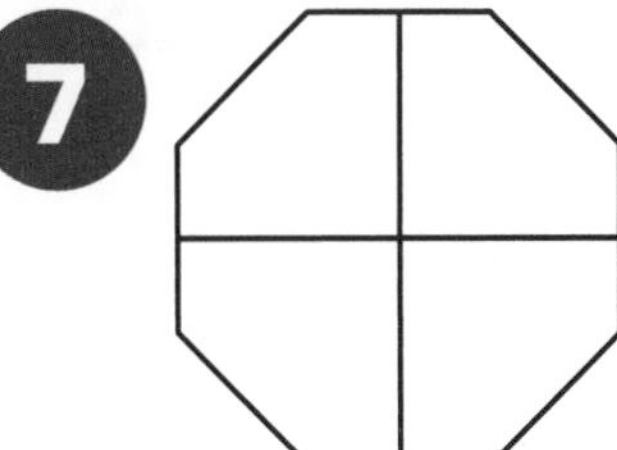

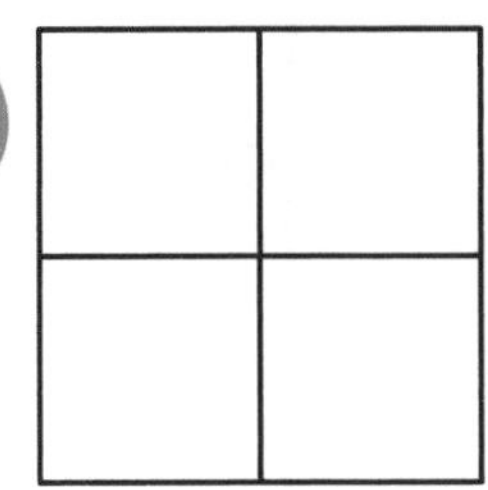

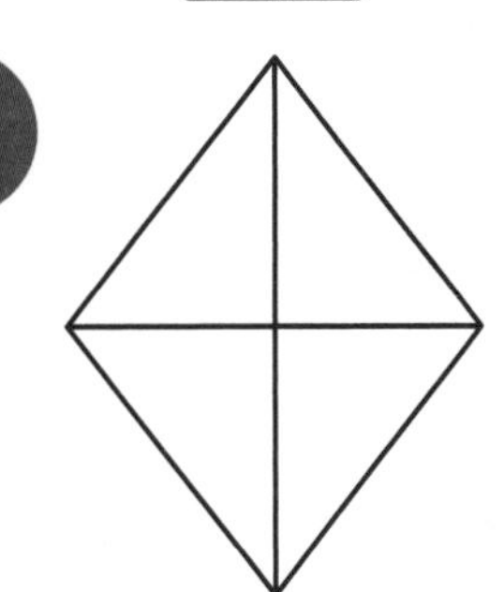

Find three quarters of the amounts.

golf balls

coins

candles

stamps

Money

You can use each coin more than once.

Use any three coins from above to make the amounts.

1 30 p
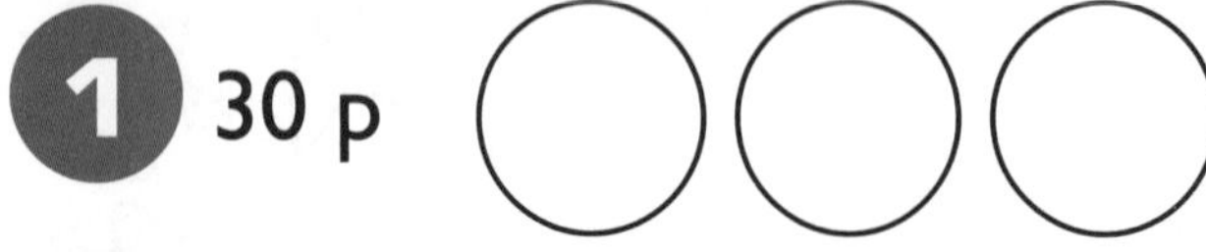

2 72 p
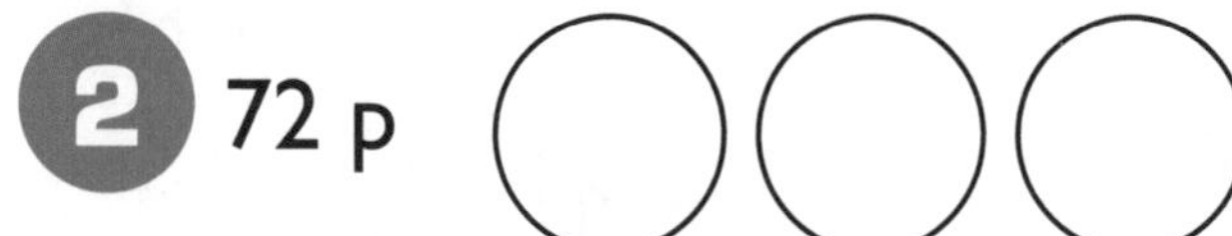

3 £1.04

4 £1.60

Find four different ways to make £2.56. Use up to six coins.

5
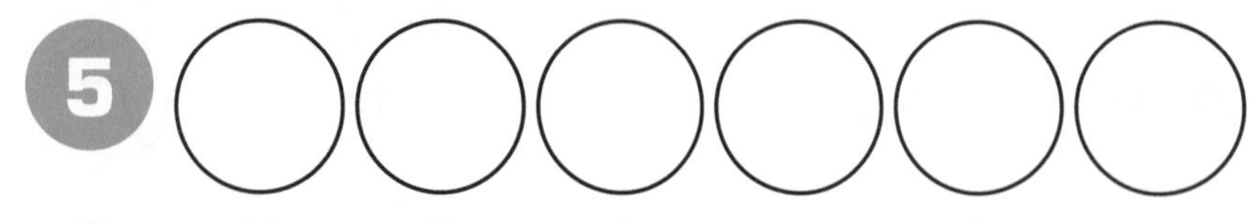

6
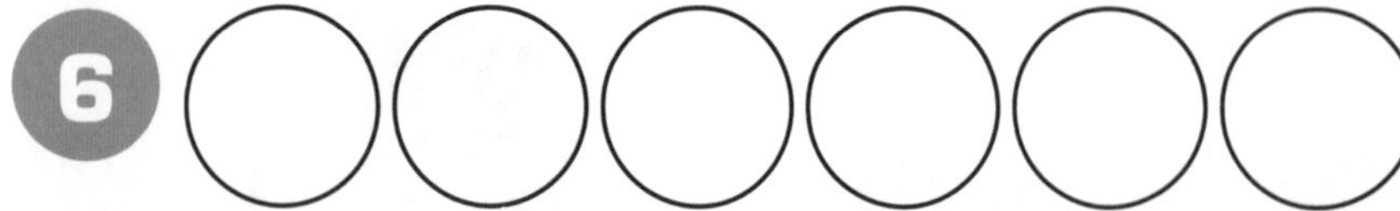

7
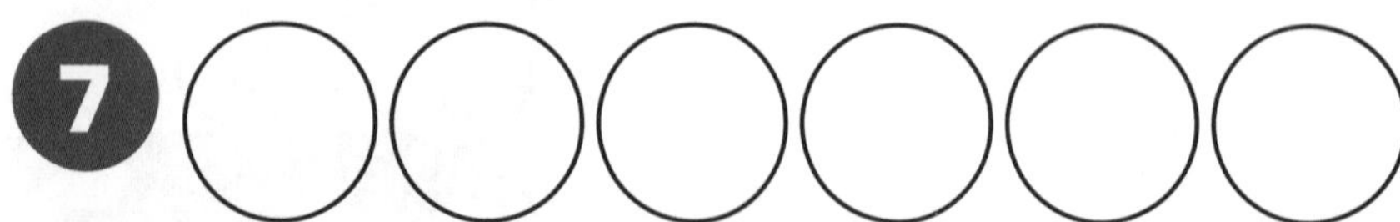

8
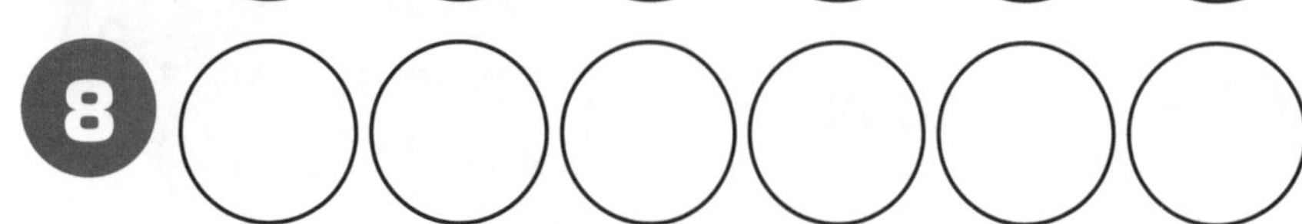

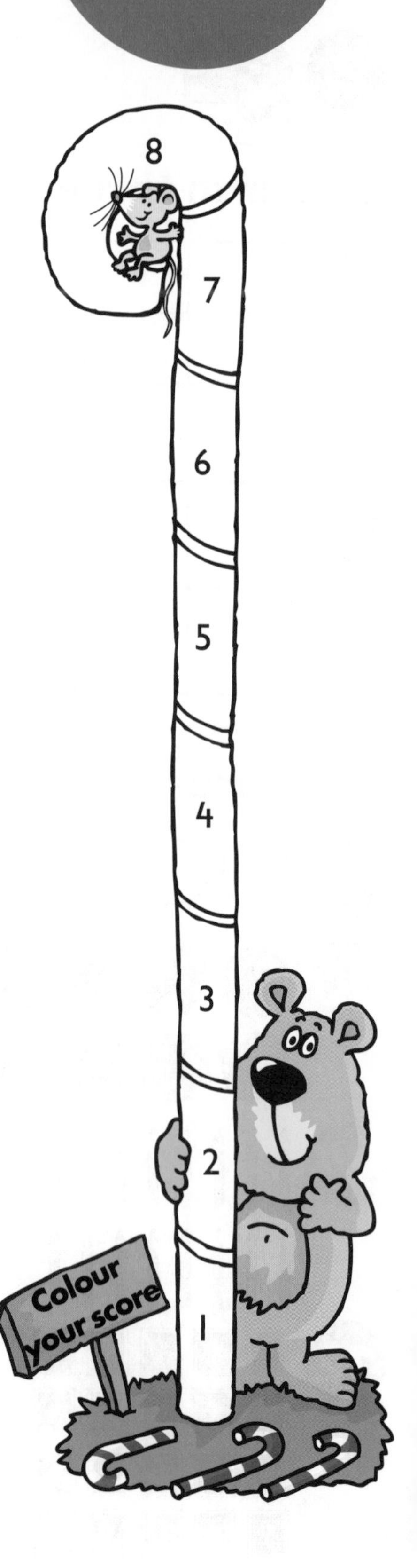

Comparing measures

Write <, > or = in the box.

< means 'is less than'
> means 'is greater than'
= means 'is equal to'

1 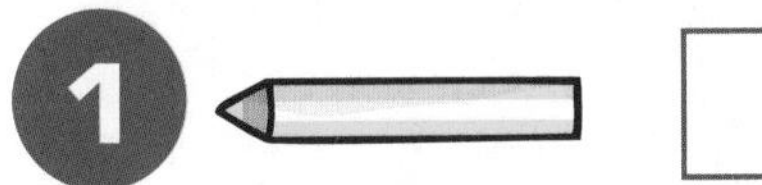 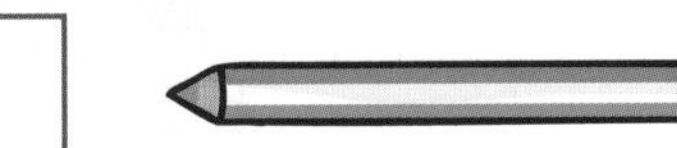

2 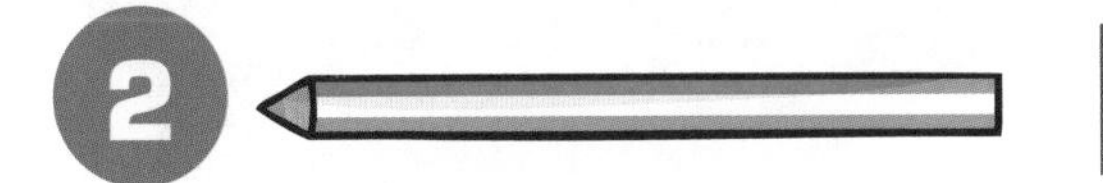  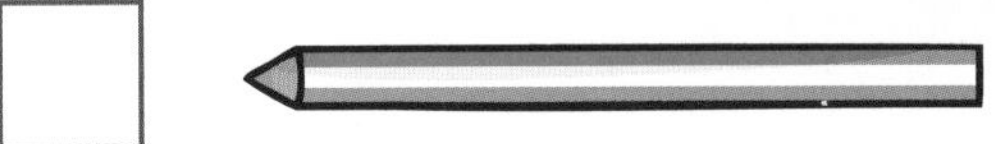

3 

4   

5   

6   

7 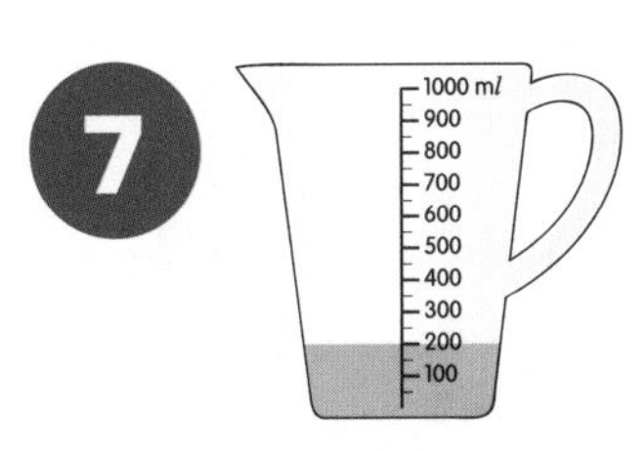 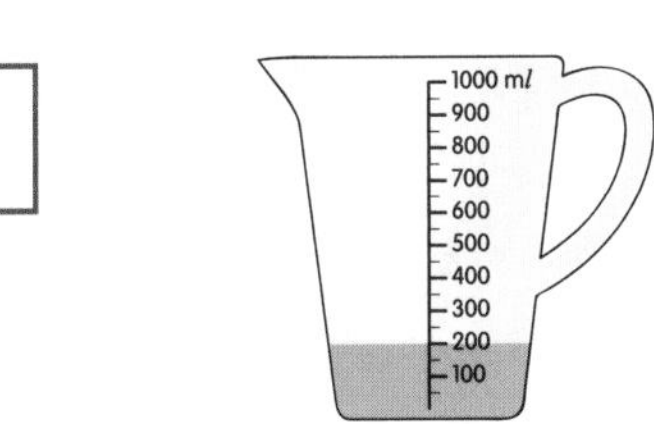

8   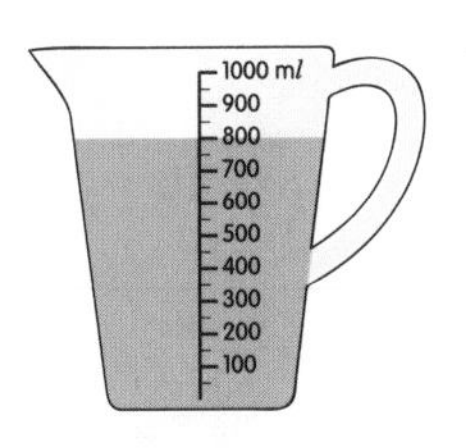

More telling the time

Write the time in words.

For the long hand, each number on the clock is equal to five minutes.

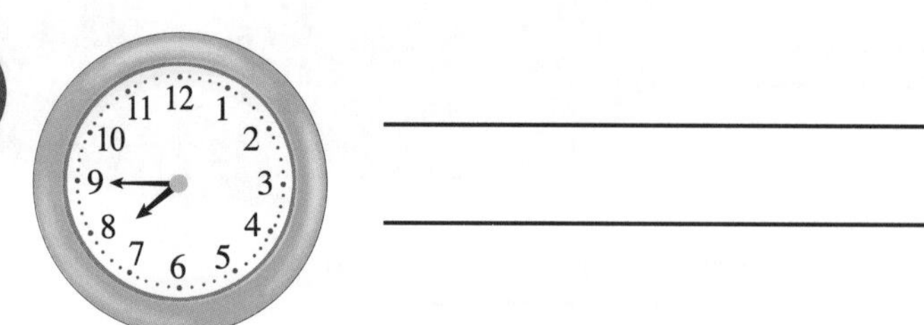

2

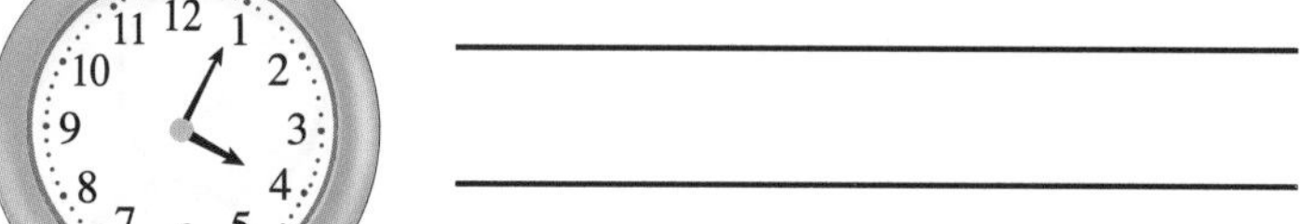

3

4

Draw the hands on the clock.

5 quarter past 8

6 5 minutes to 4

quarter to 2

Right angles

Use a tick or a cross to show if the shape has at least one right angle.

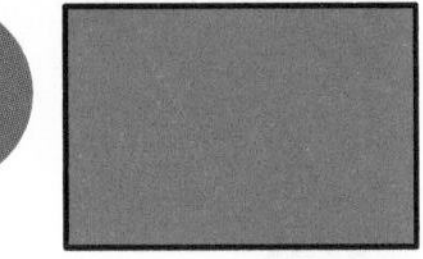

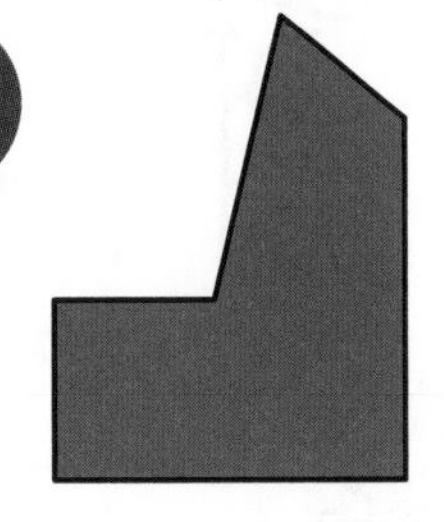

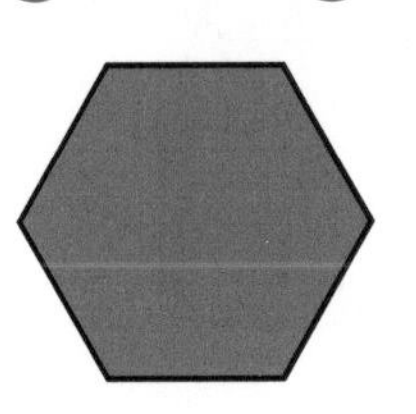

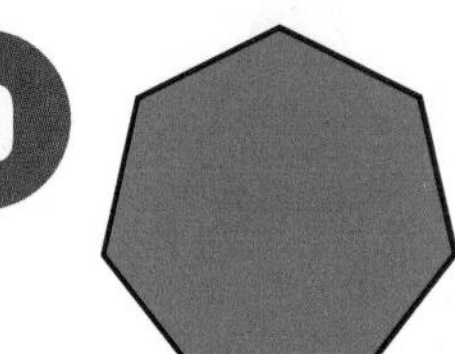

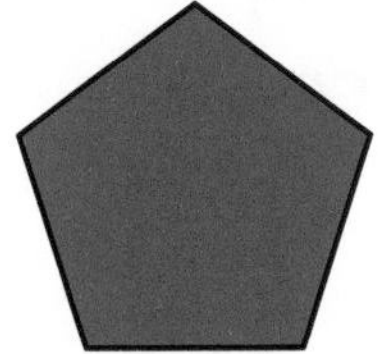

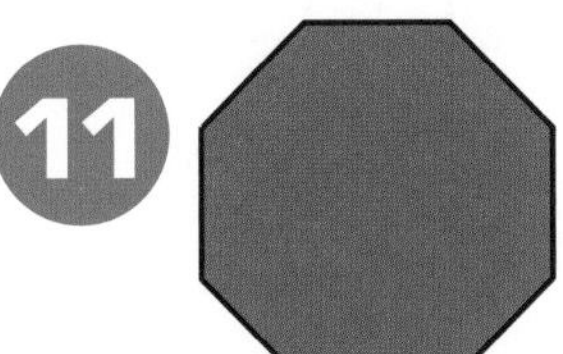

12 Write the numbers of all the shapes that have a vertical line of symmetry.

____

____

You can place a square where lines meet at a right angle.

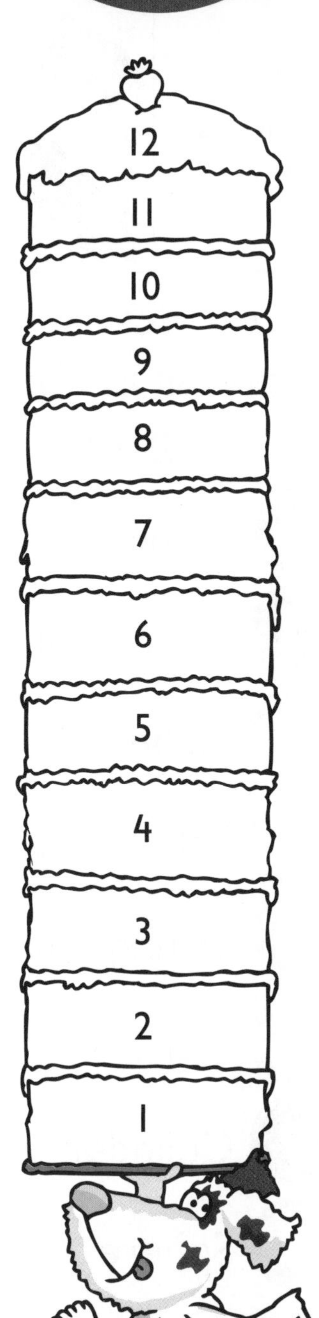

Classifying shapes

	Red	Not red
12 or more edges		
Less than 12 edges		

For each shape, check the name of the row and the column carefully.

Use a tick or a cross to show if each shape is in the correct place on the diagram.

1. Cone ☐
2. Triangular prism ☐
3. Cube ☐
4. Square-based pyramid ☐
5. Triangular-based pyramid ☐
6. Cuboid ☐
7. Cylinder ☐
8. Sphere ☐
9. Pentagonal prism ☐
10. Hexagonal prism ☐

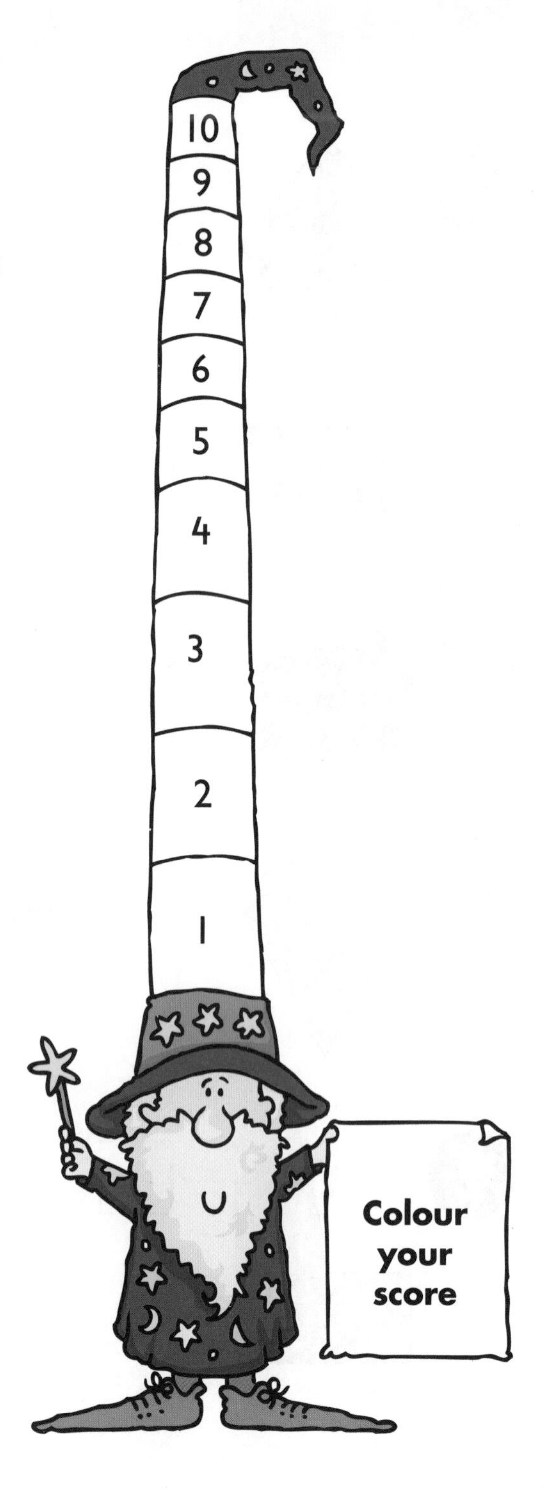

Pictograms

A pictogram to show Class Y3T's favourite physical activities

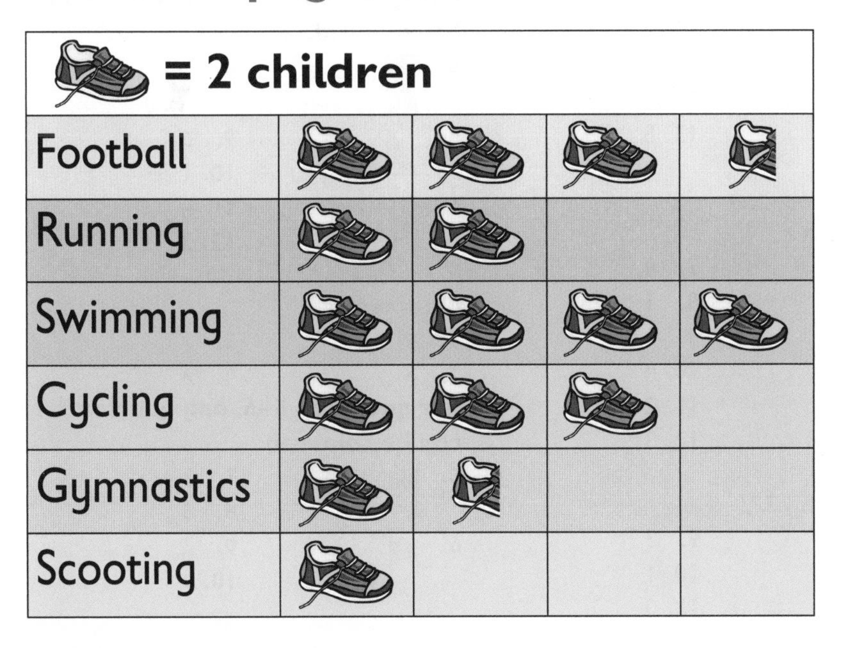

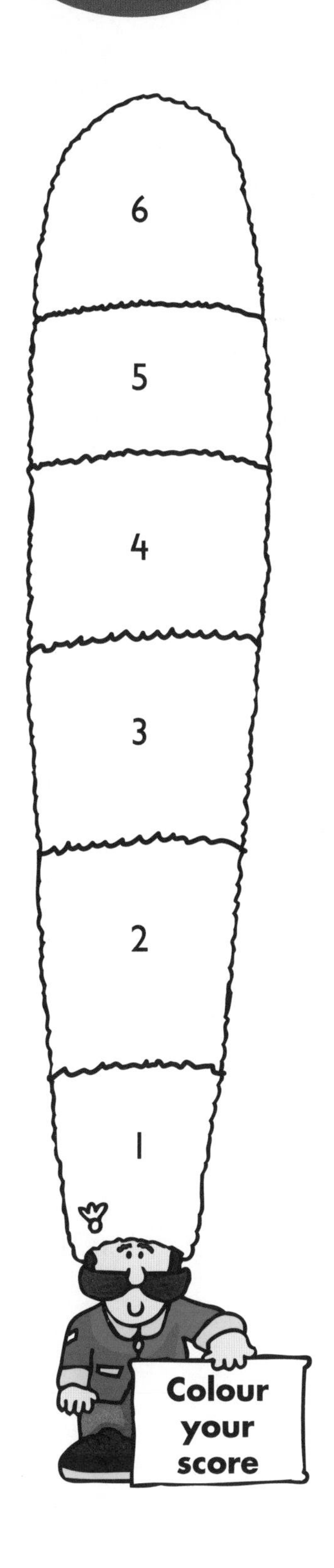

Write the answer.

1. How many children prefer football?
2. How many more children chose football than gymnastics?
3. How many children prefer physical activities that need equipment with wheels?
4. What is the least popular physical activity?
5. How many children are there in the class?
6. How many children did not choose football?

Answers

Number bonds to 20

1. 10
2. 20
3. 20
4. 10
5. 20
6. 1
7. 2
8. 13
9. 14
10. 16
11. 2
12. 9
13. 12
14. 12
15. 19

Addition

1. 9
2. 11
3. 13
4. 14
5. 7
6. 9
7. 19
8. 12
9. 8
10. 19
11. 6
12. 7
13. 13
14. 1
15. 18

Find the difference

1. 7
2. 8
3. 10
4. 5
5. 1
6. 4
7. 4
8. 5
9. 12
10. 11
11. 4
12. 12
13. 14
14. 15
15. 19

Subtraction

1. 23
2. 32
3. 46
4. 51
5. 62
6. 36
7. 44
8. 52
9. 63
10. 75
11. 14
12. 29
13. 32
14. 48
15. 58

2 times table

1. 4 p
2. 6 p
3. 8 p
4. 10 p
5. 12 p
6. 6
7. 12
8. 2
9. 22
10. 16
11. 2
12. 4
13. 8
14. 10
15. 12

5 times table

1. 10 p
2. 15 p
3. 20 p
4. 25 p
5. 30 p
6. 15
7. 60
8. 5
9. 55
10. 40
11. 4
12. 1
13. 6
14. 11
15. 8

Sharing

1. 1
2. 2
3. 3
4. 4
5. 2
6. 3
7. 4
8. 1
9. 1
10. 4
11. 2
12. 3

Dividing by 2

1. 1
2. 2
3. 3
4. 4
5. 5
6. 4
7. 2
8. 5
9. 3
10. 1
11. 9
12. 6
13. 10
14. 8
15. 7

Estimating

Accept one number either side of these answers.

1. 2
2. 8
3. 6
4. 7
5. 18
6. 14
7. 9
8. 6
9. 13
10. 22
11. 30
12. 39

Comparing numbers

1. <
2. >
3. <
4. >
5. >
6. any number bigger than 35
7. any number smaller than 41
8. any number bigger than 47
9. any number smaller than 55
10. any number bigger than 64
11. <
12. =
13. <
14. <
15. =

Halves

1. ✓
2. ✗
3. ✓
4. ✗

For questions 5–8, either half can be coloured.

5.
6.
7.
8.
9. 6
10. 1
11. 4
12. 3

Quarters

1. ✓
2. ✗
3. ✓
4. ✗

For questions 5–6, any quarter can be coloured.

5.
6.
7. 5
8. 3
9. 2
10. 4

Height and length

1. 2 cm
2. 6 cm
3. 7 cm
4. 1 cm
5. 8 cm
6. 3 cm
7. 2 cm
8. 1 cm
9. 4 cm
10. 6 cm

Telling the time

1. 5 o'clock
2. 11 o'clock
3. half past 12
4. half past 10
5. 
7. 
6. / 8. 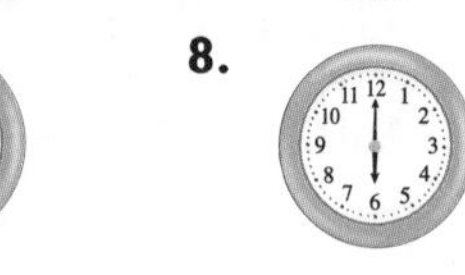

Naming shapes

1. circle
2. triangle
3. square
4. rectangle
5. pentagon
6. hexagon
7. octagon
8. kite
9. sphere
10. cuboid
11. cube
12. cylinder